Hills Newell Dwight

Foretokens of Immortality

Studies for the Hour when the Immortal Hope Burns Low in the Heart

Hills Newell Dwight

Foretokens of Immortality
Studies for the Hour when the Immortal Hope Burns Low in the Heart

ISBN/EAN: 9783337720889

Printed in Europe, USA, Canada, Australia, Japan

Cover: Foto ©berggeist007 / pixelio.de

More available books at **www.hansebooks.com**

Foretokens of Immortality

Studies "for the hour when the
immortal hope burns low
in the heart"

Newell Dwight Hillis
Author of "A Man's Value to Society"

❧

Chicago, New York, Toronto
Fleming H. Revell Company

To My
Father and Mother

THE FOREWORD

Our age is busily engaged in restating its fundamental faiths. Already the essential truths of Christianity have received new form and simpler setting. If formerly the attitude of science was not favorable to Christianity, now reasons are not wanting for the belief that faith in the great central truths of the Christian religion is steadily waxing. How the advance of physical science has affected the faith in a future life is a problem that has been discussed elsewhere. Without attempting to review that argument, I offer here these brief practical studies for " the hour when the immortal hope burns low in the heart."

N. D. H.

CONTENTS

FOREGLEAMS OF IMMORTALITY

"The faith of immortality depends on a sense of it begotten, not on an argument of it concluded."—*Bushnell.*

"I cannot believe, and cannot be brought to believe, that the purpose of our creation is fulfilled by our short existence here. To me the existence of another world is a necessary supplement of this, to adjust its inequalities, and imbue it with moral significance."
—*Thurlow Weed.*

"And hear at times a sentinel
Who moves about from place to place,
And whispers to the worlds of space,
In the deep night, that all is well.

"And all is well, though faith and form
Be sundered in the night of fear ;
Well roars the storm to those that hear
A deeper voice across the storm."
            —"*In Memoriam.*"

"It can hardly be gain for us to die, until it is Christ for us to live."—*Bascom.*

"For tho' from out our bourne of Time and Place
The flood may bear me far,
I hope to see my Pilot face to face
When I have crossed the bar."
            —*Tennyson.*

"Let not your heart be troubled. In my Father's house are many mansions. Because I live ye shall live also."—*John 14.*

Foregleams of Immortality

SCIENCE makes much of the climatic changes that have befallen our planet. It tells us that Labrador, the land of ice and snow, was once a tropic realm, a wilderness of fruits and flowers. But some disturbance gave our earth a new inclination toward the sun, and rays that had been perpendicular and powerful became slanting and feeble. Then a chill stole into the air, and the land that had never known frost was soon sheeted o'er with snow and ice, while the Amazon, hitherto the home of the iceberg, passed into warmth and perpetual summer. The climatic change that has passed over the physical world may well interpret for man the larger fact that the soul stands in a new relation to death and dying, so that summer reigns where once winter ruled.

Be the reasons what they may, all must con-

fess that society approaches the topic of immortality with a new interest and spirit.
Paganism is perishing. The old philosophy
gave us images of the scythe, the skull and
crossbones, caused the tomb to drip with
horrors, taught men to darken the windows,
to blacken the hearse, the house and the human body with plumes plucked from the wings
of midnight. But the philosophy that pictured
death as a monster, is itself death-struck and
dying. Science, that once clipped the wings
of faith, is now learning to soar and sing. If
land is not yet sighted, we sail through a summer sea, midst drifting boughs whose leaves
have not yet withered; the birds that fly over-
head belong to climes near, though still unseen; the air, laden with perfume, foretells the
continent that lies before and lures us on. Let
us, with Lowell, confess that Death, once dis-
guised as an executioner, has dropped the iron
mask and stands revealed as an angel in dis-
guise—God's seraph, come for man's release
and convoy.

Of immortality, the seer said, "We know in
part." But this annunciation of ignorance

does not destroy hope; it rather stirs expectancy. How meager the civilization that a child can comprehend! How scant the science that a babe can master! Each artist pupil fronting some masterpiece hears a still small voice, saying, "Knowledge is partial." The sum of man's wisdom represents but the merest handful. Man understands according to the nature of his faculties. The fool says in his heart, "There is no God," and rightly so; there is none—for a fool. Wisdom is not discerned by foolishness, nor music by deafness. The melody is one-half in the singer's voice; the other half is in the cultured ear. Beauty is but half canvas; its complement is the refined vision. As the Italian king, dwelling in his gloomy fortress, opened up windows through which he looked out on lakes and vineyards and distant mountains, so each new knowledge is a new window opened up in the soul's mansion through which it looks out on realms divine. The savage dwelling in the forest is like a man sitting in a dungeon, whose only outlook upon the landscape is through narrow slits in the stone wall. Sitting there in darkness the savage starves to death. But open up in him taste and imagination, and all the arts, useful and beautiful, pass before his rejoicing sight. Open up reason and memory, and he looks out upon the upward progress of society, beholds man's conflicts and victories, and like Burke, gleaning midst the ripe fields, takes on a rich, strong manhood. With each new inch added to the diameter of the telescope millions of starry worlds rush into sight. Thus with each new endowment for the soul, new ranges of the world of truth and beauty pass before man's vision. Here man carries very imperfect instruments for knowing. He is but a seed, to be grave-planted. That would be but an ignoble immortality and an impoverished futurity that man could understand. Even the poet, the sage and the seer discern but hints and gleams of the infinite truth and beauty awaiting all. Happily for us, the future of our race is inconceivably beyond anything man can discern by the utmost strength of reason or imagination.

Man is a bi-world creature. Certain flowers are biennial; they grow all summer, but the falling snows find no sign of blossom. Carry these bulbs over winter, transplant them in the spring, and the second summer will reveal their real nature and beauty. It is the universal testimony of reason and experience that life is too short for man's unfolding. Threescore years and ten are barely sufficient for developing skill in the carriage of the body. This life avails for the drill of reason, memory and judgment; all the rest of man's forty and more faculties must wait. The climate here is too unfriendly and the summers too short for their unfolding. But from the noblest specimens of the great and good we may gain some faint intimation of their possibilities hereafter: Shakespeare, disclosing fruition of reason only germinal in others; Webster or Burke, indicating the skill with which all are to think and speak; Howard and Livingstone and Lincoln, revealing the heroism possible to all. But the vast majority end their career disappointed, marred, mutilated, defeated. Shelley makes the multitudes to be "shipwrecked into life." Others are also shipwrecked out of existence. Fulfilling such a career, men are buoyed up on

11

the hope of immortality. A distinguished statesman has said: "Take away from society the belief in a personal immortality, and it may be doubted whether free institutions would survive two centuries."

When Paul recalled his career of suffering and the years during which he had been mobbed, stoned and flogged through life, he said that "were there no hope beyond he would be of all men the most miserable." But that which was a personal fact in his experience is generic, holding true of all the race. Nothing so cheap as man. The millions live upon the edge of want. Dying, each tool and place is spoken for. Two mouths wait hungrily for every crust. Our race blunders and struggles through its career. Life is like the march of an army; it is attended by tremendous losses, many falling through heat and thirst, many through sickness and exhaustion. Untold millions die having fashioned no tool, perfected no law, added no incitement to virtue. Sir Walter Scott's last entry in his journal reads: "We slept reasonably, but on the next morning——" Thus death breaks off the sentence of man's career. How incomplete and pathetic would life be for the millions without immortality! Society holds on its way despite life's defeats and pains and glooms because it believes that " on the next morning " it will enter into eternal light and infinite love.

Jesus Christ everywhere assumes immortality. He seems deeply conscious of the great over-world; it ever lies before His mind like a noble landscape familiar and seen since birth. By reason of that vision splendid, men toiling below the sky seemed to Him only laborious triflers. But what was unconscious knowledge with Him comes to us by slow processes. From the day when Socrates and Plato and Cicero defended their faith in immortality it has been the duty of each man to marshal his arguments and set in order his proofs. Christianity asks no man to take anything for granted. Everything is to be tested in reason's crucible.

Passing in review, therefore, the arguments of philosophers, poets and seers with those of Jesus Christ, the reflective mind notes the presumptions, structural and constitutional, in man. Consider the fact of self-identity. You are the same person you were twenty years ago. Yet the physiologists insist that you have had several bodies during this score of years. Even the bone system has been thrice replaced. But life taxes the brain so severely that its fibre is replaced twice each year   Recalling the events of to-day and yesterday, men also recall the faces, landscapes and events of three-score years ago. A distinguished lawyer once said that while in the midst of an argument, and under great mental excitement, there began to rise before his mind the pages of a legal decision he had read thirty years before. Slowly the dim and misty lines grew clear; at length he read them with perfect distinctness. Surely, in this event memory was no physical scar   Doubtless that which was unique in his experience exists in germ form in us all. The "I" gives unity to our knowledges and experiences. Self-identity gathers up all past life. Having survived the changes of many brains and half a score of physical bodies, the soul begins to nourish the hope that it may survive the body altogether, casting it off like a worn-out garment.

Reason finds a foretoken of immortality in the contrast between the growth of material things and that of the mind. When a tree fulfills leafage, flowers and fruit, it touches the limit of its being. Its end is fulfilled; growth can achieve nothing more. But all this has its absolute contradiction in the mind. Not even of the ripest scholar can it be said that reason has touched its limit and exhausted its capacity. Contrariwise, each new discovery, each new invention, does but prophesy other achievements and nobler acquirements. To-day's goal is only the starting-point for a new journey to-morrow. Nor can it escape our thought that an ever-growing vine or oak would be an infinite calamity. Channing says: "One tree endowed with unlimited expansion would come to overshadow the nations, exclude every other shrub, and exhaust the world's fertility." For reasons of utility, therefore, the growth of birds and beasts and shrubs must be limited. But no such necessity holds the mind. Burke can take all knowledge for his province and in nowise exclude Webster. Michael Angelo was equally great as architect, sculptor, artist and

15

poet. Yet each new intellectual achievement did but stimulate and nourish the great minds about him. Macaulay at last developed such skill in acquiring languages that in six weeks he mastered Italian, read its great writers, and began his critique on Dante. Similarly, Sir William Jones mastered thirty languages, and thought a single three months sufficient for a new dialect. The more the mind acquires the more it can acquire. Nature says to all: "To him that hath it shall be given." Each new form of thought or virtue or philanthropy does but double the possibilities for self and others. When the mind goes abroad for surveying the universe it comes back with the reflection that the world was built for limiting and ending the body, but for continuing and forever nourishing the mind with the heart and conscience.

Socrates was the first to find immortality in a certain indestructibility in things. Smite as man may, he can destroy nothing. The coal burns, but its ash and smoke precisely equal the original bulk. Each petal dropping from the flower does but make the next rose redder.

Trees fall, but do not perish; they only take on different forms. Fire, wind and water have whipped and lashed the atoms from one end of the universe to the other. Full oft these poor particles may have desired to lie down and die, but this boon was forbidden. To-day there is not an atom less than at the world's beginning. Now, Watt's thought, that builds an engine, is inconceivably greater than the raw iron it organized. Yet the mind exhales ideas as the sun emits light. Should all man's thoughts be recorded there would be a volume for each day, a shelf filled for each month, a library for each year. Is all this heart treasure to perish when atoms are made to persist? No sage nor scientist can give his mental treasure over to his child through heredity. The seer's wisdom perishes with him, and his babe must begin where its father did—at nothing. Nor is there any racial immortality. The time comes when our sun will be a burned-out cinder and our planet a dead world. Does God care for atoms, and make trees abide though their forms change, but bring the heart that laughs and weeps and loves and aspires to that end called a black hole in the ground? God-led into the world, the soul is God-guided out of the world. To live again to-morrow is no more wonderful than to have lived at all yesterday. In regnant hours the soul scorns proofs and despises arguments and exultingly sings: "God is; therefore I shall be, forevermore."

A certain irritating force in man seems to foretell immortality. All growth is through a kind of hidden stimulus. When growth begins, restlessness overtakes the child. Quietude becomes impossible. To condemn the little creature to a chair is a form of exquisite cruelty. Nature needs fresh blood in the extremities for her building processes, and secures it by pricking the child until it runs and jumps. When the period of growth is passed the restlessness also passes. Similarly, after periods of sickness, with convalescence comes restlessness, and this restlessness compels the exercise needed for health and strength. But with old age comes quietude; the easy-chair foretells the end.

A like irritation exists in the mind. A noble discontent inaugurates each new epoch for man.

The soul chafes against its barriers. Sometimes this world seems like a tiny garret on a hot August night, and the heart will smother unless it finds breathing-room in a larger world. In his dungeon in London Tower Sir Walter Raleigh could pace but twice his length. Thus the soul cries out against the limits of a dungeon life bounded by these walls called the cradle and the grave. It asks all the air there is between itself and God's throne; it needs this room. It wants all the sweep between God's throne and the eternities; it is to move in this orbit. As the child's restlessness stimulates exercise, growth and maturity, so the aspirations of the heart are preparations for and prophecies of an immortal destiny.

To-day scientists are interpreting anew the instincts in animals and men. Instincts are nature's prophecies foretelling coming events. Through them animals guard against possible danger and attain happiness and maturity. The spider in California builds a large passage way to its nest, and, opening therefrom, a secret passage with a trapdoor. Similarly each insect and bird carries some like instinct for nature's needs; and instinct never deceives its possessor. When the young lark commits itself to the soft air, the receiving medium always bears it gently up, while the robin's migratory instinct always finds the southern clime foretold. We must also reckon with that faculty in man looking forward to immortality. In vain we ransack all nature for a single instance in which nature's instincts have deceived insect or bird. Does nature use so great skill for guiding beasts, but become a blunderer in guiding man? Nay, further: Does nature, through instinct of what awaits them, speak truth to wasps and spiders and sparrows, but tell lies to man of what awaits him? If man lives for this world only, then God is become a mere purveyor for the body. A mother is justified in the disagreeable tasks related to her infant by the foresight of what the babe will become. And if man is to drop his body and hereafter develop his rational faculties, God's care for the body is justified and explained by His foresight of the excellence to which the mind and heart shall attain when time and infinite resources have accom·

plished their fruition upon the soul. But if man's life is limited to the body, then the Creator is reduced to an infinite cook and racial restaurateur. This theory says that there is a God, but that he is a fool God, exhausting all his resources in growing pumpkins and potatoes for the inside and wool and flax for the outside of man's body. This is thinking become stupidity and brutishness; it is intellectually as absurd as it is morally monstrous. Science is rapidly reducing doubt of immortality into sheer mental vacuity. Our planet is one. When the traveler crosses the Hudson the laws of light and heat and gravity are the same on the New York side as on that of New England. God's moral universe is also one. Man's life here and there is hemispheric. Crossing that stream called Death, man moves forward under the embrace of the laws of intelligence, self-consciousness and freedom.

From the presumptions of immortality structural in man, reason moves to a higher plane of argument, and marks the actual beginnings of resurrection in the individual. Fundamentally, man is mind entombed in flesh.

His reason is embodied in the grave-clothes of matter. Doubtless the grand climacteric fact is the complete shaking off of the body. But every form of conflict with appetite and passion, in which the spirit throws off its bondage to the flesh and becomes victorious, is itself a partial resurrection. When, after long entombment, the tulip bursts forth into its full blossom, it attains its perfection; but every form of secret growth that split off the outer bulb and pushed forth the stalk and leaf, was a part and prophecy of the final floral outburst. Very pitiful oftentimes the struggles of men who wrestle with themselves and battle against their evil tendencies as for life itself. Every man is double, and the lower man grips at the very throat of the higher man and spiritual. Jugglers bound hand and foot know how to shrink their muscles and slip out of the rope or fetter. But man is not so skillful in escaping his physical thongs. Now and then we see a man who, after all the thunder of life's battle, stands in life's evening light victorious over himself, with nearly all the evil brood of appetites and passions slain within him. The

22

partial conquest of these animal forces here
foretells the sublime emergence from the body
hereafter. Each victory now is a morning star
foretelling the rising sun. God gives partial
resurrections here as foretokens and first fruits
of the greater resurrection at death and beyond it.

The climacteric resurrection in death has its
forerunner in the sufferings that refine gross-
ness out of man and redeem him out of a low
life into a higher. Plato thought the single
suffering with its exalting uses a foretoken
of that great death-suffering that forever
ends all ignorance and sorrow. Pains are
blows of the hammer knocking off the
rough outside of the geode to release
the beauteous crystals within. Troubles are
blows lifted upon the dungeon door for giving
the prisoner release. Sufferings are stamp-
mills crushing the quartz that the gold may be
free. Looking forward to the fruit, men plant
the peach seed. But a thick shell entombs the
living germ, nor has the shining of the sun any
power for letting the plantlet out. Winter
alone can resurrect the little life out of its seed-

grave. Therefore cold drives the frost wedges into the thick shell and splits it; cracking, the seed is rent apart; then the germ hears the call of the light and the air, and, rising into the realm of sunshine, sets forth upon its career. Man, too, is buried in his physical life, and suffering comes in to give release and resurrection. That which single suffering begins, the great death-suffering completes, giving life and final resurrection.

These hopes, germinal in man, burst into full blossom in Jesus Christ. In Him the intimations of nature and human life cease to be cold proofs and become an enthusiasm and a faith that comfort and satisfy the heart. What a finished statue is to the block of marble the resurrection of Christ is to the general doctrine of immortality. He wrought immortality into perfect shape. "We may know," says Theodore Munger, "that there is a statue in the marble, but how beautiful it may be, in what grace of posture it may stand, what emblems crown its head, what spirit breathes from its features, we do not know until the inspired sculptor has uncovered his ideal and brought it to the light." We see in the statue the marble, but we also see the artist's mind. So in Jesus Christ we see not only the fact of immortality, but its special meaning and possibility. Nature's proofs and foretokens are cold. Nature is beautiful, but has no sympathy. She gives her flowers alike to bride and to bier. When the poet Lowell's heart was well-nigh broken with grief, it came to him with a great shock that nature had no care for him. "Not a bee stints its humming, the sun shines, the leaves glisten, the cock-crow comes from the distance, and yet but a moment before the most immediate presence of God of which we can conceive was filling the whole chamber and opening arms to suffer the little one to come unto Him." But in Jesus Christ God sympathizes; God sees; God cares. Before Him the earth ceaselessly exhales spirits into the heavens as the sea its white clouds. But going, all move under His convoy and divine love. Disappearing, they do not die. As they who sail over the seas go down into the vessel and for a time disappear from sight, so has it been said, "the grave hides for a little time, but does not destroy." Thus the grave is the shutting of angels' hands, that they may safely keep the treasure and convey it to the other side. Jesus Christ is the soul's door opening into immortality.

For each Christian heart the Easter morn is full of sacred memories, and also has its pledges and suggestions. What possibilities does it unveil ! Soon the hidden man shall be released from the body, that seat of pain and disease; that Circe's palace where angels dwell, and also demons; where passions glide darkly, and also strike; where vices nest and lusts have secret lairs. The soul entombed in a sluggish, obese temperament is like an angel doomed to draw a plowshare. Therefore a thousand congratulations to those who have cast off the clog and look out of life with winged thoughts. Their crowning achievements here are but twilight intimations of the mind's creative force there. Even the most notable natures, like Plato, Milton, or Bacon, do but suggest the vastness and volume of thought possible to all under higher conditions. It is only when we select the superior minds with the strongest faculty in each—the poet, the thinker, the philosopher, the universal genius in execution—and melt all these glorious gifts into one new and nobler being, that we perceive the full - orbed man who is to come. What possibilities, too, of friendship and affection does immortality open up! No force for good like personal force. Here, indeed, man must be guarded against and parried; here men strike, men pursue, men blight, men destroy. But what if each man stood over against his fellow for stimulus, balm, and bounty? What if drawing near to a friend was like approaching a star that blazes and sparkles with ten thousand effects? How rude is friendship here; how stunted is human love! It grows only as oak or evergreen in arctic regions. There the tree rises but a few inches from the ground, but carried south it springs full two hundred feet into the heavens. Embowered there it prefigures that volume of love to which all shall some day come. To those on whom life's burdens rest heavily, defeated, despoiled, homesick for those who have gone, comes this hope of im-

27

mortality. Believe him who said "death is sweet as flowers are; death is as beautiful as a bower in June." The grave is like the gate in the old cathedral—iron on one side and beaten gold on the other. Perhaps our gravestone is a gate for those whom we have loved and lost. We say, " A man is dead;" God says, "A man lives." Dying is transformation. Dying is home-going, happiness, and the Father's House.

IMMORTALITY AND LIFE'S WITH-HELD COMPLETIONS

" There is, I know not how, in the minds of men a certain presage, as it were, of a future existence, and this takes the deepest root and is most discoverable in the greatest geniuses and most exalted souls."— *Cicero.*

" My general wish on earth has been to do my Master's will. That there is a God, all must acknowledge. I see Him in all these wondrous works. Himself how wondrous! What would be the condition of any of us if we had not the hope of immortality? What ground is there to rest upon but the Gospel? There were scattered hopes of the immortality of the soul, especially among the Jews. The Romans never reached it; the Greeks never received it. There were intimations crepuscular twilight; but, but, but God, in the Gospel of Jesus Christ, brought life and immortality to light."—*Daniel Webster, on his death-bed.*

When Rufus Choate took ship for that port where he died, a friend said: " You will be here a year hence." " Sir," said the great lawyer, "I shall be here a hundred years hence, and a thousand years hence."

" Immortality is the glorious discovery of Christianity."—*Channing.*

" The sad memories which death brings are a part of our education. Under the influence of an absent soul the heart softens, and man goes forth each day more of a friend to his race, and more of a worshiper of his God. Sorrow must ennoble duty, not end it. The death of a friend exalts those who remain to weep."—*David Swing.*

Immortality and Life's Withheld Completions

IN EVERY age the master minds have believed in immortality. For the sons of genius and liberty the soul is cosmical, not planetary. Immortality seems an infinite invitation upward. In Tennyson and Browning the spring tides of life run so deep and strong; for Emerson and Lowell life is so full of laughter and songs and sighs, so full of struggle and victory, that hope expands the handful of years into immortality. Call the roll of the great names of history, and each inspirational nature will contribute some testimony to faith, akin to Wordsworth's "Ode to Immortality." In these children of beauty and culture hope vaults forward like a rainbow into the deep future; no great poet cares one whit because the archangel's wing is not strong enough to return and report what lies at the end of hope's beauteous bow.

As those who dwell inland from the coast ever hear the muffled sound of the distant sea, so he who lingers long o'er Hamlet or Lear will hear unceasingly the waves of the infinite sea breaking upon the eternal shores. Each Dante and Milton also shows us sky rising above sky, and heaven overarching heaven, even as one star rides high above another star. Upon his raft of reason Socrates sailed down the river of life, and when the night fell and the ocean heaved dimly in the vast dark, with a tranquil face he put boldly out and sailed the sea with God alone toward that eternal continent where light is ever constant beyond earth's gloom. With like faith Plato looked forward unto that realm where earth's exiles shall be disentangled from the toils of ignorance and sin. Not even an atheistic education availed for extirpating in John Stuart Mill the faith of personal immortality. After all life's fierce conflicts with doubts and questions, a remnant hope still survives in each Greg and Mill, and this fact witnesseth to immortality far more strongly than does faith in some believing Browning. In great men immortality is reason prophesying. The hope of immortal life dies only with a dying God, just as the falling planets would mean the falling of the central sun.

Emerson profoundly says: "When the Maker of the universe has points to carry in His government He impresses His will in the structure of minds." Thus, in all the animal and vegetable world, the wish of the Creator is organized into the created. The maker of each loom or press accompanies his mechanism with a book of directions concerning the tapestry that will be woven or the pages that will be printed. In like manner we may logically infer that the divine mind will accompany each rosebush, each apple tree, each skylark, each human heart, with a handbook of directions called instincts and automatic forces. Now nature has fulfilled this expectancy. No rosebush is ever left in doubt as to whether it should bear red blossoms or thorns and thistles. To each young bird there comes a secret voice, bidding it trust its weight to untried wings and soft air. Through the boundless sky also the inner voice guides the water-fowl in certain flight. Obeying these voices, the fish swims, the bee hives its sweets, the bird builds its nest. Having never once been deceived by these secret instincts, the vegetable and animal realm attain the end of their being and fulfill their destiny. From these instincts in the creature science learns how to interpret the plan of the Creator. Each Agassiz returns from his survey of the world with the feeling that his hopes are God's written guaranties of immortality.

It is as if all nature had broken into voice and through the soul uttered her "everlasting yes." He who meets the bird's wing with air that bears it up, the fish's fin with water that yields to its movement; he who meets the eye with sunlight and beauty, the ear with sweetness and melody, hunger with bread and thirst with flowing springs, hath filled the soul also with hunger for immortal life, with thirst for eternal love. At times this hunger becomes so great that man could stretch up his hands and "eat the planets like small cakes;" his thirst is so deep that the earth itself is but a small cup for the soul to drink in. Did God give man this infinite hunger only to find afterward that his generosity had involved Him in penury, so making it impossible to furnish man with bread wherewith to satisfy his hunger? This would make the Infinite to be either poverty stricken or a moral monster. Here millions die in ignorance and millions in sin. The joy of one heart is marred by the anguish of another; the wealth and beauty of one street by the pathetic poverty and shame of another; the music of one voice is destroyed by the moans of another. But God doth tempt all men upward toward the heavenly heights with dreams of a land whose clime is eternal spring, whose air is perpetual music, whose life is endless joy. These aspirations are liens upon immortal life. They are stepping-stones that "slope through darkness up to God." Out of them science and faith are building a new heaven and a new earth.

Human life is a colossal enigma without immortality. The hypothesis of a future life alone can explain man's troubles and solve his mysteries. The inequalities of society baffle all intellects. Bad men rise to the throne, the good are forced to the wall. Tyrants dwell in kings' palaces, heroes starve in dungeons. Often vice wears purple and fine linen; sometimes virtue eats crusts and wears rags. When Dante was denied his vine and fig tree, wicked princes drove in chariots from palaces in the city to villas in the country. Why is it the heroes of liberty and religion have been hunted like partridges upon the mountains? Tiberius flung his victims over the precipice into the sea. Nero lighted up his gardens with blazing martyrs. But these tyrants lived on to the end in splendor, and died on soft rosebeds, as did the murderers of Socrates. Meanwhile, where are the patriots of liberty whose lives were one long struggle against tyranny and oppression? Where are your fathers, who sleep at Shiloh and Gettysburg, where the hillsides are all billowy with graves? What about that mound in the forests of Africa where Livingstone fell? If death ends all, what compensation had Savonarola and William the Silent and Lincoln?

The inequalities of mind and heart are greater. Oliver Twist, living in Fagin's den, his teachers thieves, his trade crime, his only education gained in the school of iniquity, gives us pause; but Oliver Twist stands for multitudes of orphan boys in every city. Our physical atmosphere is laden with soot and smoke. No statue in the park but is blackened. No picture on the wall but holds some grime. Every marble in the gallery has some black dust on the white forehead. Thus man's moral world is full of ignorance and sin. Every mind hath suffered some injury, and every heart is heavy with some pain. Trouble is big with mystery. Against its granite wall in vain we strike our black and bleeding forehead. Than Job none hath done more to solve it. If this is all, then for the multitude suicide is life's chiefest boon. But what if death brings compensation, and beyond, all wrongs may be righted? Beholding in the perfected race the fruitage of their toil, the patriot and martyr will find in their continued life the explanation of life's every ill.

If man be immortal, his ideals and aspirations, unfulfilled here, may be realized hereafter. In imagination, every plan is complete and every ideal perfect. Each purpose hangs before the mind's eye like a heavenly vision. But the ideals suffer grievously in the work of embodiment. By the time the plan has passed through man's mind and been formulated, it is crippled and sadly disfigured. Beethoven tells us his polished symphony is but an empty echo of the heavenly music he heard in his dream. The generations have gazed enraptured upon Raphael's Sistine Madonna. But the artist painted it with anxious face and left it with troubled and disappointed heart. Try as he would, the painting, as we see it, is only an attempt to reproduce the vision Raphael saw, but could not fully realize upon the canvas. What poet or prophet ever fully uttered all his dreams? What philanthropist ever realized all his reforms? What statesman ever overtook his ideals? Does not each new discovery open a thousand new and hitherto unsuspected possibilities for the inventor? Dying at ninety years of age, Humboldt was still an eager student. Feeling that he had just begun to learn how to study, the great naturalist exclaimed: "Oh, for another one hundred years!"

But what if there is another life ? If Humboldt here thought through gross nerves and brain, what if there he thinks through fine ether ? What if Beethoven has completed the chords broken and interrupted here ? What if Socrates has finished the argument interrupted by the jailer's hemlock, and justified the ways of God, to Critias ? The canvas Raphael painted has endured for three centuries. But has God ordained that the canvas shall be preserved while the artist has fallen into dust ? Is "In Memoriam" more than Tennyson ? Is St. Paul's cathedral more than Sir Christopher Wren, its architect ? Is the leaf to live, while the tree dies ? Reason and conscience whisper, it can not be. If thoughts live, the thinker can not die. To suppose that death ends all is intellectually as absurd as it is morally monstrous. Because God lives, His children shall live also.

The immortal life furnishes the explanation of the early dying of those from whom society has the right to expect the most of good. The list of illustrious ones whose star sank back to the horizon before it had approached its zenith, is long and sad. At twenty-two, Keats knew that he must die, and wrote his epitaph: "Here lies one whose name is writ in water." And Shelley, his friend, whose soul, rising, poured forth sweet notes, like the skylark of which he sang, died when only thirty. Mozart died at thirty-six; Raphael at thirty-seven; Burns before he was thirty-eight. No man in all his generation had a clearer vision, or promised more for his age, than Frederick W. Robertson. Dying at thirty-seven, the scholar-preacher exclaimed: "It is all a mystery. Man is like a candle blown out by a puff of wind." In a single week after Fort Sumter was fired upon the colleges of our land stood silent; deserted all their classrooms. When several years had passed the rooms had filled again, but not with the old students !

Pathetic, also, is the death of the ten-talent minds, to fame and fortune all unknown. Several years ago some one clipped a little poem from an obscure country paper and sent it to a great magazine. Scholars read it with delight. An inquiry for its author was instituted. This poem, bearing the mark of genius, proved to have been written by a boy—a section hand upon the railway; and he was dead. This we know—no more. Why did the harp break after the first song was sung? Why died that noble boy, Arthur Hallam, whose genius promised so much for English literature? Young Charles Emerson rose above Harvard College like a rising sun. If that sun perished when it disappeared, what signified its rising? Angels have entered our homes—"their footprints graves." Departing with them have gone our dear ones who were best fitted to live.

The "In Memoriam" reminds us that the "forbidden builders" are a great multitude. With long life, from them there was nothing, nothing we might not have expected. Gladstone is 88, but his voice hath not lost its charm. Pope Leo is 87, and his mind still hath its cunning. Bismarck is 82, but his iron will and purpose are still potent. Had Robertson and Shelley and Arthur Hallam and all these children of genius lived to 80 and beheld the golden setting of life's sun, what treasure might have come to our generation!

If death is all, then is folly chargeable upon
the universe, But if life goes on beyond the
grave, if these royally endowed ones continue
their creative work under new and higher conditions, if there Raphael's best work awaits our
admiring vision, if Keats and Charles Emerson
are singing there, then physical death ceases
to be an absurdity and becomes the highest
wisdom. With Tennyson let us believe that
the task, incomplete here, will be completed
hereafter by "the divinely gifted man."

The withheld completions of life also ask for
immortality. Life is full of wrecks and failures.
The march of a generation is like the march
of Alaric's army leaving their cold forest home.
The forest children turned their faces toward
the sunny land of Rome. In the forward march
of the Teutons, boys and girls fell by the way-
side, overcome by heat; parents fell through
hunger and exhaustion; wounded soldiers were
constantly dropping out of the column, to
die in the thicket, unmissed and uncared for.
Thus there are myriads who end their career
having lived indeed three score years and
ten without having fulfilled life. The larva

eats its way out of its cradle, consumes the leaves upon the bough and falls back into dust. Thus whole tropic races live only an animal life, using the mind to gain supplies for the body. They were born, they ate, they died—this is their history. With others failure is ancestral. A hundred years ago the weapon that wounded them started upon its way. Many, through a single mistake, have wrecked life and happiness. A traveler crossing some mountain pass may indeed fulfill a thousand right steps, but by one hour of carelessness slip upon the precipice and henceforth go crippled, and by one error some have overthrown a thousand deeds of uprightness.

Others pass their early life in districts remote from knowledge and wisdom, and only in middle life discover their power. These rude and undeveloped ones are like geodes—outwardly they are rude and rough, inwardly they hold flashing crystals. Some end their career wholly unrecognized by those who walk through life beside them. As in the western prairies men plow and plant their harvests

over veins of coal lying hidden and unsuspected; as in far-western mountains men build
their cabins over hidden veins that hold fabu-
lous wealth were they discovered, so some men
complete life and fall on death never knowing,
never dreaming that talent and skill were
latent in them, like undigged treasure.

Some there are who, if this life ends all,
are indeed most miserable, in that they have
involved in grievous suffering those dearest
to them. How pathetic these moral incomple-
tions! When the thunder-bolt smites the tree
in the forest, it also blackens the beauteous
vines and flowers that wrap it round.
Sometimes when society visits its scorching
penalties upon the wrongdoer it also smites
the innocent mother or wife or child.

The most piteous part of those letters that
come from Mexico and foreign climes, whither
men have fled from the consequences of their
evil deeds, is their consciousness of suffering
brought upon the innocent wife or mother.
That in injuring himself, the wrongdoer has
blighted other lives, lends agony to agony
itself, adds poignancy to deepest pain. And

oh! with what desire do such men beseech God for an opportunity of retrieving their errors and sins ! In the best, goodness is only germinal. Men go toward death stored with latent faculties and forces, just as our winter-bound earth goes toward May—stored with myriad germs and seeds, waiting for summer to unlock and send them forth to bud and blossom and fruitage. There are unexplored riches in the human constitution. What is man ? No one knows. Many of his faculties exist in him like unwrapped tools in a box—not even examined, much less named. Three or four of his forty faculties ask threescore years for development—the other latent powers ask an immortal life for growth beyond the grave.

If, therefore, death ends all, life is robbed of its dignity and deeper meanings. Man spends seventy years toiling upon his industries, his arts, his books, his friendships. Each achievement in character is a victory after a fierce battle. In our world wisdom never comes unasked, and no virtue stays unurged. Character in man is like beauty in the statue—it asks for infinite pains. But no Phidias would toil unceasingly upon his Parthenon if he knew that once the peerless temple was completed the destroyer's hand would pull it down, leaving not one stone upon another. Praxiteles would hardly have carved his matchless Venus, fronting the certainty that when the finishing touches had been given some enemy would lift the hammer and break the precious marble into a thousand fragments. Our generation builds its libraries and galleries, its temples of science and religion, in the hope of permanency, and sends these structures down as heritages to coming generations, even as Henry VII. sent Westminster Abbey down to the London of our day.

Thus also the highest motives for culture and character come from the thought of permanency and personal existence in a future life. Good men cannot abide the thought that, dying, they will be unwelcome and unknown when they enter the presence of the patriots and heroes, the brave and true and great of yesterday. The thought that character achieved not only lends happiness here, but happiness and worth hereafter, supports them in the long, fierce conflict with ignorance and sin. Dignity and honor can hardly attach to him who journeys forward toward a black hole in the ground. In view of the difficulties that confronted Clay and Garfield, in view of their days of poverty, their nights of study and struggle, the harsh winds that assailed their bark; in view of the fact that when death overtook them they had scarcely begun to work out their dreams, it seems difficult to believe that the brief and fragmentary success achieved in this life was worthy the heavy price they paid.

We joyfully confess there is more happiness in virtue than in vice, in culture than in ignorance. But if man builds a house just in time to die and have his body carried out of it; if he gives himself to unceasing study, to find that threescore and ten years avail only for gathering a single handful of flowers from each garden, a single cluster from each of earth's many vineyards; if he founds a business only to discover that the outlay of strength means that the business must pass into other hands; if he loves and is loved, binding and being bound with hooks of steel, only to find that all dear ones must be torn from his arms—if this is all, is man more than the insect of an afternoon? Has life's game been worth the candle? It is the immortal realm that lends life its exalted meanings and messages. Let us believe with Tennyson, that man is supported here by the hope that "life shall live forevermore."

To-day science is uniting with faith to strengthen the argument for immortality. Gone forever the age when science denies the future life! No scholar is more distinguished than Professor Pope, who says: "He who believes personal immortality is unscientific believes on insufficient evidence." In view of physical phenomena as yet unaccounted for; in view of thought transference, mental suggestion and telepathy, the great scientists of all countries scorn the expression, "Brain secretes thought as the liver secretes bile," and hold that as now the mind uses brain and nerve, it may later on use ether.

Call the roll of the great chemists, physicists

and biologists of Germany and England, and almost without exception they are on record as teaching that death does not end all. The preparation of this vast world-house, its adornment and furnishing by millions of years of preparatory work, the development under divine guidance of man's intellectual and spiritual forces to the end only that man may live an average of three-and-thirty years, turns the universe into a riddle without any meaning. Has the world-architect and artist toiled for nothing? Is man ephemeral, "a bubble that bursts, a vision that fades?" A thousand times nay! answers that new science represented by John Fiske. "I believe in the immortality of the soul," says the scholar, "not in the sense in which I believe in the demonstrable truths of science, but as a supreme act of faith in the reasonableness of God's work."

The old skeptical science is becoming obsolete. Atheism has gone into bankruptcy again. Nature has ceased to be a rival of God. The universe is only the physical body through which God works. Nature as a self-sufficing mechanism has disappeared. The old science emphasized physical facts and forces. The new science says that the idea is a fact as truly as a paving-stone. Drop a stone out of a high window, and it kills the man on whom it falls. But Garrison dropped from his window a tiny piece of paper on which were written Christ's words, "Do unto others as you would have others do unto you," and that idea fell upon slavery with the force of ten thousand earthquakes and ground it into powder.

During the American Revolution, when the English Secretary of War urged an increase of troops in Boston until their guns outnumbered the guns of the Americans, Pitt is reported to have said: "We must reckon not so much with their guns as with their sentiments of liberty." The great statesman knew that not rifle-balls, but sentiments, win battles. And the new science perceives that instincts and aspirations in the mind are facts of nature that must be interpreted and accounted for by reason, as truly as a stone in the hand.

That oration beginning, "Life is a narrow

vale between the cold and barren peaks of two
eternities; we cry aloud, the only answer is the
echo of our wailing cry," represents in popular
form the old scientific scholarship. But the
new science laughs at this crude culture. Only
very ignorant people can any longer say, "We
do not know, because no one ever came back."
No wide-spreading, acre-covering oak of 200
years ever put off its gianthood, folded up its
stature and went back to an infant acorn. No
Webster ever became an infant a second time
and re-entered the cradle that he might whisper to the new-born babe the experiences of
his superb physical manhood. Nature whispers to each babe, "The sage and seer may
not return to you, but you may go to them."
Nature whispers to each artist pupil, "Your
Master may not return to you, but you may go
to his side." In all the realm of field and
forest, of land and sea and sky, there has never
been a ripe, mature thing that has returned to
unripeness in order to become the instructor
of crudeness and rawness. If a thousand
statesmen, dying, returned to unfold free
institutions to a babe, the going back would

51

avail nothing. The babe must go up to Lincoln's level to understand his message. Only
a sage can understand a sage! Only a seer
can understand a seer! Man must go up add-
ing sense to sense, and faculty to faculty, in
order to attain scientific demonstration of the
future life.

From these intimations of immortality nature asks each to weave in one cumulative argument a chain that may not be broken. Mathe-
matical proof may not be possible for every
mind. That man should live again is not so
strange as that man should live at all. Now that
the steam engine has been invented it is easy
to foretell its continuance. Newton's mind is
more than the clod beneath his feet. But Neptune is only many clods brought together. It
is inconceivable that the great God grants an
orbit of millions of years to that wintry clime
and clod called Neptune, but gives Newton,
the philosopher, whose mind squeezes the planet
for truth as the hand squeezes an orange, a
career of but threescore years and ten. In-
credible the thought that God makes the
"Principia" to endure, but permits its author
to fall into dust.

When a benefactor has bestowed a thousand favors upon some youth, and so carries himself as to imply another gift, it would be an act of supreme meanness to doubt the continued kindness of the benefactor. The noble mind and the generous heart will trust God for the larger hope. God lures the soul forward by filling it with dreams of a land where rude speech has become eloquence; where the misshapen face gives place to lustrous beauty; where the one-talent man shall go on toward supremest genius; where, like the tree of life, each mind shall bear fruit every month; where music is marred by no discord; where all love all and all serve all; where life means growth, power, maturity, beauty; where the sorrows and woes of hero and patriot and parent shall hang on the walls of memory like the shields of vanquished enemies. This is the immortal life and the eternal love that Christ hath brought to light. As at the northern cape the midnight sun sinks below the horizon only to flash up again in the dawn of a new day, so man dies that he may live again. Man is God's child. Man is immortal, because God is eternal.

CHRIST AND IMMORTALITY

"There can no evil befall a good man, whether he be alive or dead."—*Socrates.*

"Where I listen, music; and where I tend, bliss forever."—*Browning.*

"I came from God, and I am going back to God, and I won't have any gaps of death in the middle of my life."—*George McDonald.*

"I am not afraid to die, but I wish I might carry on my work. I have only half used the powers God gave me."—*Theodore Parker, on his death-bed, to Francis Power Cobbe.*

> "Ah Christ, that it were best
> For one short hour to see
> The souls we loved, that they might tell us
> What and where they be."
> —*Tennyson.*

"The prize is noble and the venture is great."—*Plato.*

"The body of Benjamin Franklin (like the cover of an old book, its contents torn out, and stripped of its leather and gilding) lies here food for the worms; yet the work itself shall not be lost, for it will, as he believes, appear once more in a new and more beautiful edition, corrected and amended by the Author."—*Franklin's epitaph, written by himself.*

"I know that my Redeemer liveth."—*Job xix. 25.*

CHRIST AND IMMORTALITY

WHEN the modern student opens his Cicero he is depressed by the gloom that lies upon the pages of the orator and scholar. Accustomed to a sunny literature, the modern thinker marvels that the fear of dying and death makes up so large a part of Cicero's private soliloquy and public writing. The Roman lawyer was the first citizen of his day. He was the child of genius and of great wisdom also. His were honors, public and private; his was wealth, with the splendor of the city residence and the beauty of the country villa; his the friendships of the great. Yet, when his beloved daughter Tullia died a gloom fell upon the statesman that did never again lift. Overcome with grief, Cicero denied himself to all friends and retired to the seclusion of his Tusculum villa. But when he opened his books his favorite authors failed him—he could not

see to read for the flood of falling tears
When Socrates' fearlessness of death and
Plato's arguments for immortality had failed
to console him, then Cicero prepared his own
arguments for immortal life and a readjustment beyond the grave. But, alas! neither
literature nor time availed to heal his broken
heart. The charm had fled from the arbor, the
glory had gone from landscape and sky, from
the favorite fountain and the leafy woods. For
the daughter, Tullia, Death had stolen all
beauty from the cheek and made her marble
brow to be of clay. But for the father, Death
now dimmed the color of gold, withered the
wreath of fame, made empty the orator's ambition. Gone forever the zest of life! Each
task seemed unworthy its toil. "Death may
come to-day," said Cicero. "It is always
hanging over us like the stone over Tantalus."

Among the ancient worthies not Cicero alone
suffered a tragic career through fear of dying
and death. Recent excavations in the lands of
the Parthenon and of the Pyramids tell us that
the civilization of the ancients clustered about
the tomb rather than about the temple. What

a revelation of ancient life is found in **the** grottoes of the dead uncovered in **Mycene** and Thebes! The coin, the bronze urns, **the** terra cotta, the priceless manuscripts there found, tell us that treasures denied **the** living were freely given to the dead. Often-times a hovel was the home of the living members of the family, but for the dead was prepared a polished palace. There, in the sculptured sepulcher, after the farewell funeral feast, the dead were enclosed, their weapons by their side, the provisions made ready for the journey into the unknown fields, the works of some favorite poet lying just at hand—pathetic proofs these of the immortal hope that burned low, indeed, yet has always burned upon the altar of the human heart. But each bas-relief, each stately poem, each page of the philosopher, was stained black by the terror of dying and death Even Homer makes Achilles, newly returned from the shades, to say he "had rather be the meanest slave on earth than king among the dead." Art also shared in this degrading fear. So far from the lily being the symbol of death, its emblem was the skull and crossbones, its color was black, its music was the dirge. Looking upon some happy group gathered by the fireside, Death beheld the scene as so much ripe grain, and made sharp the sickle. Not only did death lend fear to man's daily life, but its cold chill stole into literature also. It fell like a black bar across the sunny pages of each Cicero.

Doubtless the danger of violent death was responsible for some of the gloom of ancient literature. In an age when tyrants and despots flourished, the philosopher, the statesman and the poet were constant invitations for the headsman's ax. In that far-off time each king had his hired poisoner, his paid assassins. Of all that goodly company of great men of whom Plutarch wrote, how few died a natural death ! Socrates was the crowning glory of Athenian civilization. Yet the city fathers voted to kill their first citizen with a cup of poison. Cæsar was at once general, statesman, orator, author. One day when he was planning to polish his writings into classic form, the daggers of the senators stilled his voice and pen. Thrice Cicero was the savior of his country's liberty. Grown gray in the service of the people, he hoped to complete yet another book. One afternoon soldiers from Antony entered his garden with a written order for his execution. "Strike me," said Cicero, "if you think it is right." A minute later the head of Rome's greatest orator was lying upon the ground. In an age when all public speakers and writers were liable to brutal execution, men naturally spoke and wrote much upon the gloom of dying and death. But for whatever reasons, ancient books, ancient art, ancient poems, ancient life, all unite in the confession that through the fear of death men were all their lifetime subject to bondage.

Looking back to the era when the world was in its morning time of letters and life, and noting that for some reason the solemn and threatening voice of tragedy had become the prominent note of ancient culture, Macaulay has observed that while Christianity has changed the face of Europe and won a thousand triumphs, "its crowning glory is that it has wiped the tears from eyes which had failed with wakefulness and sorrow, lent celestial visions to those dwelling under thatched roofs, and shed victorious tranquillity upon those who have seen the shades of death closing around them." How different the influence of that sepulcher digged in Cicero's villa and that tomb opened up in Joseph's garden ! If the first brought the eclipse of every joy to the orator, Christ looked forward to His death as the hour when, having been obscured for three-and-thirty years, His sun should break through all clouds and shine forth in untroubled splendor. Great, indeed, the influence of Tullia upon Cicero's life and thought, yet the disciples suffered a thousandfold greater loss in losing their Master. He found them friendless fishermen. Taking them to His fellowship, upon them He poured all the treasures of earth's rarest and most glorious friendship. He found them dull and low-flying, and gave them wings and aspirations. He found them cold and frigid, and lent them warmth and inspiration. Daily in His presence what had been latent, and in germ, unfolded into bud and fruit and flower. After His arrest, His trial and pitiless execution, they fled away cowards, skulking into the darkness for hiding-places. Yet let us confess that something happened within a few days that affected these men morally and intellectually in some such a way as liquid iron is affected when it is hardened into the strength of unyielding steel. Some influence transformed these feeblings into men of oak and rock, and freed them from bondage to the fear of death. If Plato argued, the note of conviction stole into the disciples speech. If Cicero had doubted, certainty crept into their affirmations. If Demosthenes the orator and Socrates the philosopher had received death as a necessity, these men began to woo death as a friend. Freed from the old conceptions that made the grave to drip with horrors, that counted death an executioner, these men welcomed death as a messenger, wearing indeed an iron mask without, but having within the face of an angel of God. Each disciple therefore was ambitious to achieve a violent death. Paul looked toward the heads-man's ax in Rome; James was hurled from the battlements in Jerusalem; Jude was slain by the mob; Philip was hanged upon the scaffold;

Peter was crucified in Persia; James the less was killed in Asia; Matthew was slain in Abyssinia. They smiled to receive the ascending flames, as other men smiled to receive the robe of scarlet, or the golden cross that makes the knight. Seeking to account for the fact that in three centuries Christianity had achieved the throne of the Cæsars, the Roman skeptic declared these disciples conquered through the new view of immortality, that released them from the fear of dying and of death.

Confessedly, Christianity's view of immortality immediately enhanced the sense of individual worth. Not until value attaches to man himself will his law, his literature, his art, his institutions take on value. If in our age man has grown humane, so that drinking-fountains are prepared for dogs, and the state lifts the shield above the horse, protecting it against the cruel scourge, in that far-off time human life itself was inconceivably cheap and worthless. Ten thousand men were slain in the Coliseum during the reign of a single emperor. A philosopher no less eloquent than

Cicero defended the awful spectacle. When citizens of the better class gave a banquet the entertainment was not considered complete without a sword-fight that left half a dozen slaves dead in the presence of the assembled guests. Even Pliny praises the husband who celebrated the funeral of his wife with one of these bloody spectacles. The modern Siddons or Salvini would have failed utterly to please those audiences, accustomed to nothing less than the sight of mangled bodies. In view of the fact that medieval Europe was constructed out of the fragments of ancient Rome, a scholar has suggested that when the fifteenth century made the fagots or the wheel or the rack ready for heretics, that age was only continuing those Roman games which gave such delight before the despotism of politics became the despotism of religion. In such an age the miserable and the unfortunate were numbered by legions. The enemies of happiness were so numerous and so powerful that suicide became a popular resort. In the morning the general put on his short sword, so that in the event of bad news from the army he might fall upon his weapon. Also the citizen thrust his dagger into his belt. If the day brought bad fortune, a way of escaping the sorrows of the forum or the market-place was always near at hand.

In such an age slaves opened the furrow, slaves carried the sheaves into the shocks, slaves had charge of the wine press, slaves also quarried the marble; under the master's direction slaves carved the statue. The city, with its homes and streets, its fountains, its parks, represented the toil of slaves. The master lay upon the couch while the slave wrote down his thoughts. Later on he rested from the grievous toil of listening to a beautiful poem, while the slave went for some ice to cool his lordship's spiced wine. Under such conditions master and slave alike suffered an inconceivable degradation. The justice and self-respect that make civilization great were threatened with utter destruction. The rights of the master were keenly felt; the rights of the man were undreamed of. When, then, it was asserted that the soul was immortal, that each slave bore two worlds in his heart, that in a second life the master's injustice and cruelty would have its reward, and the bravery and moral courage of each Epictetus their rich recompense, a great change began to be felt. It was as if spring had released the icy fetters of winter. A soft warmth stole away the rigor of cruelty. Iron laws became gentle. Since the wrongs done here would be righted hereafter, the scourge fell from the hands of the despot. Looking unto that immortal shore, men saw the flag of equality unfurled above prince and peasant alike. There all outer trappings were seen to have fallen away. He who here was on the throne and robed in purple was there, perhaps, seen to have been abominable. He who here dwelt in a hovel and knew neglect, there, perhaps, stood nearest unto God's angels. Immortality greatly enhanced the sense of individual worth.

Christ's idea of immortality also made a powerful assault upon the vices of society. No age has been unacquainted with immorality, yet during the age of Cicero the generations went thundering into evil courses like hordes of wild beasts, unrestrained and irrestrainable. The pictures upon clay tablets and the bronze vases found in Pompeii and Herculaneum tell us that in those cities vices were once worshiped that now make horrible the very name of Sodom and Gomorrah. In temples also sins were crowned with chaplets of flowers that are now outlawed by society. When his master tied Epictetus to a post, and with a lever twisted his leg and made the sufferer a cripple for life, no one thought of punishing the rich man for his cruelty to the slave, who also was one of earth's great philosophers. Even in Plato's day, in the city of art and eloquence, mothers exposed their children under the law by which "he who claimed the child might hold it as a slave forever." When Seneca and Lucian affirm that virtue was unknown in the Roman empire of their day they tell us that vice and sin had injured the cottage and palace alike, as the storm never can injure the storehouse and granary. The generation that believed death ended all went rioting through life, trampling down sweetness and innocence as the wild boar tramples down rosebuds or lifts its tusks upon the perfumed shrubs. But when Christ asserted that the good men do lives after them, that God will bring every work into judgment with every secret thing, that in the day of revealing, in the presence of neighbors and kindred, every deed is to be traced through society as a seed is traced to its wide-spreading harvest, then vice and sin were assaulted in their secret strongholds. From that hour immorality began to wane. Vices hitherto recognized became matters of public shame. Crimes that had crawled like serpents through the streets of the city were either scotched or killed. The fact that of all the ancient vices drunkenness and social evil alone have come down to our day is a powerful argument for the influence of Christ's idea of immortality exerted upon the vices of society.

The immortal hope has also strengthened man against the woes and wrongs of life. In every age man has known misfortune. Blights blast man's harvests, storms wreck his ships, his house burns up, his bridge falls down, his laws are imperfect, his rulers are corrupt, gratitude has failed, and at last old age overcomes strength. By reason of the increased comforts and conveniences our age, by way of contrast with that of Cicero, may be called "the age of universal happiness." But in that era, because man's ignorance was great, his unhappiness was great also. Society suffered woes many and grievous. Not infrequently a newly appointed general signalized the event by raising a company of soldiers and sailing away to some distant province. In that far-off land his soldiers went forth to farm out the taxes. They ravaged the village and the farmhouse, they swept the very land for concealed treasure. When one of these generals returned home he brought with him riches so vast as to support a series of entertainments of which a single dinner cost a modern fortune. Going home at midnight each guest was accompanied by a slave who carried the goblet or the dishes that appealed to the admiration of the friend. In such an age the people suffered many woes and wrongs. Poverty was extreme. The average family, it is believed, had an income of only thirty dollars a year. The common people also

owned no land. For the most part they dwelt
in hovels. Their clothing was of coarse hemp.
He who broke a bone must go through life
with a crooked limb. The ignorance of sani-
tation involved fevers and epidemics destruc-
tive beyond all present knowledge or belief.
Taxes were exorbitant. Freedom from war
was almost unknown. Governments also were
cruel. The courts were the instrument of oppression for the strong. The soldiers fell un-
noticed in the forest, the sailor sank unknelled
into the troubled seas. The good and the wise
after long lives of nobleness entered into pain
and weariness and oppression through despots.
We need not wonder that, dwelling amidst
such conditions, the early writers tell us that
had men believed that they died as the beasts
do society would have broken down under its
weight of trouble. But if men bore up under
their woes—toiled on in the hope of ultimately
righting all wrongs and curing all social ills,
sought to exchange ignorance for the arts, and
coarseness for noble manners, and achieve progress for society—they did all this through
the foresight of that realm " where the wicked

cease from troubling and the weary are at rest."

It was this faith in the immortal life also that gave the heroes their conquering courage, the reformers their immortal renown. History would be robbed of half its splendor without the story of the patriots and martyrs who have endured, seeing afar off the life and the land that are invisible. That man who was stoned at Lystra, mobbed at Philippi, beaten with rods at Iconium; who endured perils at home and perils abroad; who faced a world in arms, and at last, with dilating form and kindling face and with "the diapason of the sea mingling with his speech like noble music unto noble words," cried out, "None of these things move me," achieved his matchless fame as a hero through the foresight of immortal life awaiting him. Also the realm invisible supported the Waldenses, the Huguenots and the Puritans. What if bloody Nero lived in a golden house, while Paul was chained in the dungeon of the Mamertine prison! What if Lorenzo dwelt in a palace and wore purple, while Savonarola dwelt in a garret and ate crusts! What if the cruel English Queen did make soft and silken her nest, while her executioners were drying the fagots for Cranmer in his Oxford jail. Beyond, every wrong would be righted, and there full justice would be done. If the enemy stills the looms of action here, the threads shall be taken up hereafter! If here the patriot is lifted upon the cross of slander, there the truth shall be fully known! Because there is a readjustment beyond, heroic souls stand out like the rock of Gibraltar midst a sea of troubles. Sir Galahad, the knight, asked for the hardest task and petitioned for the place of greatest danger, for he anticipated the hour when the knights should return and, assembled around King Arthur's round table, should rehearse the deeds of heroism, and receive from the King's hand the just recompense of reward.

If to-day, midst the din and whirl of life, society has come to emphasize the inner manhood and character rather than the outer and bodily conditions, the new estimate of worth is due to the immortal outlook. Since beyond, character is the all important thing, here also for the sake of personal manhood men have

made great sacrifices. Urged by his friends
to polish his writings into perfect form, a
great scholar exclaimed, "My books and culture can wait until that second life," and so
went on serving men. "Ease can wait,"
said Xavier, toiling for the ignorant. "Pleas-
ure can wait," said Macdonald, toiling in the
tenement-house district. "Leisure and comfort can wait," said Arnold Toynbee, as he
served the helpless. "Luxury can wait,"
saith a great company, who deny eye and ear
and outer sense that they may fulfill the
higher duty. In the coming realm character alone is of priceless worth. It is
the foresight of that revealing day that re-
strains avarice and enterprise, that rebukes
ambition and the pride of honor.

We return from our outlook with the thought
that the vision of the new heaven has made for
man a new earth. The light falling from the
heavenly shore hath lent a soft radiance to
man's earthly life and thought. Handel tells
us that when he wrote the "Hallelujah Cho-
rus" he saw the heavens opened and all the
angels and the great God himself. When death

robbed Tennyson of Hallam, his friend, the poet took up the harp of life and, looking toward the immortal realm, music of unwonted sweetness stole over the world. Dying at last, he passed away to the music of his own requiem. But the vision splendid hath not simply lent a new sweetness to music. Because man is to live again, he hath hastened to double his culture and purify it, to double his art and refine it, to ennoble his laws, to expel coarseness from his literature and make it divinely beautiful. The immortal outlook has given man all great art, all great work, all great character. For man goes singing, weeping, aspiring, praying through life, journeying not toward a grave in the grass, but toward a statelier Eden. When the little child, the sweet mother, the poet or statesman falls asleep, should we look up with Dante we would see "a divine chariot sweeping through the heavenly confines, its pathway well-nigh choked with flowers."

THE WITNESS OF GREAT MEN TO IMMORTALITY.

" For love is stronger than death."

" Love has never denied death, and death will not
deny love."—*H. M. Alden.*

" I go to prove my soul;
    I see my way as birds their trackless way.
    I shall arrive! What time, what circuit first,
    I ask not; but unless God sends his hail
    Or blinding fireballs, sleet or stifling snow,
    In good time, his good time, I shall arrive.
    He guides me and the bird. In his good time."
                                        —*Browning.*

" Never the spirit was born the spirit will cease to
        be never;
    Never was time it was not; End and Beginning are
        dreams !
    Birthless and deathless and changeless remaineth
        the spirit forever;
    Death hath not touched it at all, dead though the
        house of it seems ! "

THE WITNESS OF GREAT MEN TO IMMORTALITY.

IN ALL ages reflective minds have brooded long over the concealments of Nature and the silence of God. Clouds and darkness surround God's throne, indeed, but the throne from which man doth rule is also girt about with silence and mystery. Having carefully concealed man's origin and made obscure the beginnings of his thought language and morals, Nature has passed on to make thick the clouds about his tomb. How amazing the fact that the poet and the dramatist, in portraying the procession of life, must make the sweet mother and her babe; all lovers, with their youth and beauty; all scholars, with their learning; the hero and the patriot, to hasten forward into that breathless and worldless mystery called the Realm of Death!

Behind walls of granite Nature seems to hide herself. Some secrets man hath, indeed, succeeded in wresting from her unwilling hand. He hath found out how to make sweet the fruit which Nature makes bitter; how to make hard metals soft; how to hew marble into a temple; how to make a desert become a garden or a city. But neither tears nor prayers nor longings have availed to wring from the lips of Nature the awful secret that hangs above man's grave. The stone castles, within which kings and emperors secrete themselves, will fall like houses of pasteboard before man's heavy cannon. Nor can the banker devise a safety-vault of steel that man's sharp chisel can not penetrate. But Death can build a wall that defies attack. Though the doors of the grave, at the touch of an infant's hand, swing inward to receive the newcomer, a giant's hand cannot cause the adamantine door to swing outward for his release. All feet, whether walking in the paths of glory or the paths of obscurity, are journeying toward a grave digged midst the waving grass. Even lovers who clasp hands in an eternal friendship shall soon discover that their embrace was an eternal farewell. The happy parent, who in the morning swings the gleeful child, shall, when the sun hath set, find his hot tears falling upon a little face grown cold in death: for the streets of every city converge toward one point—the old churchyard. If Youth whispers, "Man is fashioned like a god," Age echoes, "Man fades like a leaf."

Nevertheless, there is a silence that thunders. Nature hath a concealment which is revelation. Secrets there are that proclaim them· selves upon the housetops, and life hath a hori· zon that speaks eloquently of a continent, hidden, indeed, but real. For shallowness alone hath no secrets. It is superficiality that tells the full story. But the unseen forces, the chemists in the roots, the monarchs of the clouds, the giant forces in the harvests—these work in secrecy and silence. Gravity doth not blow a trumpet before it. The sunbeam doth not lift up its voice and cry aloud in the street. If summer, journeying northward, drives the arctic winds back into their icy caverns, summer's loudest tone is the soft whisper of the south-wind.

For Nature's silence is only seeming; her concealments are big with testimony. Every apple blossom blushes forth its secret of the rosy apple that is to be. Every acorn throbs with the germ of an acre-covering oak. Every seed aches with its thoughts of a golden sheaf that soon will ripen. The perturbations that deflect Uranus from its path proclaim the new planet soon to stand upon the horizon. The great discoverer tells us that at the very darkest moment of his voyage he received overtures from the unseen continent. The ocean currents bore golden branches upon their bosom, while through the air came the land-birds—the birds of paradise, brilliant with color—and, pouring forth their thrilling songs, welcomed Columbus to a continent hidden, indeed, beyond the horizon, but a continent that was so great as to involve apparent concealments of distant rivers and valleys, of forests and mines and mountains. How small the continent that could in a single day have revealed itself to the discoverer!

Just as sailors, when they are still far out at sea, know that they are drawing near home by reason of the odors of shores as yet unseen; as

The Witness of Great Men to Immortality

Sir Launfal, after long years of absence, stayed his tired horse beneath an old tree many miles from home, yet heard the tones of the bells in the old abbey sending sweet welcome on before; as in that picture called "The Aurora" the watchman in the night saw the feet of the dawn standing upon the mountain-tops a full half-hour before the sun rose in the sky—so, if clouds are about man's tomb and silence above his grave, for him also there are rifts in the clouds, there are moments when heavy draperies of darkness part, there are voices that fall softly through the air, whispering that man's home is not the tomb on which we strew flowers and shed tears

Man's thoughts outnumber the sands; his hopes exceed the stars. Sometimes his tears hold a sorrow that is deeper than the sea—his friendship, who can measure ? Life that is so rich, so sad, so happy, ought also to be long. When little Roland, sitting upon the knee of King Charlemagne, besought the great ruler to tell him what treasures were to be his, the wise King shook his head and grew thoughtful How could a little child receive from a great ruler the full story of castles and palaces and kingly realms that were to be his when childhood had widened to the measure of these great treasures? In that hour Charlemagne's silence and concealment foretold Roland's future wealth. Thus the mystery and silence about man's tomb encourage hope, not tears. That God, who can cause his sun, from a point ninety millions of miles away, to release the world from its tomb of ice and its shroud of snow, and clothe that world with forests and shrubs and flowers, is fully equal to the task of lifting man out of the winter of death into the summer of immortal life.

"There is," said Cicero, "in the mind of man a certain presentiment of immortality; and this takes deepest root, and is most discoverable, in the greatest geniuses and the most exalted souls." Professor Tyndal! also tells us that in his higher moods the faith of immortal life was strong in him, and that only in his sodden hours did it fade away. And Wordsworth, recalling the years when health was perfect, and his young heart pure and innocent, said: "Meadow, grove and stream, the earth and every common sight did seem appareled in celestial light." By "the vision splendid" the young poet "was on his way attended." At last, when the cares of this world increased upon him, the glory faded away, the vision died in the light of common day, and the things which he had seen, Wordsworth saw no more.

Ours is a world where, as the telescope weakens, the stars die out of the sky. As sublimity was partly in Mount Blanc and partly in the mind of the adoring Coleridge; as loveliness was partly in the daisy and partly in the mind of Burns, who saw it—so the star of immortal hope rises for him alone who hath eyes to see, and the voices divine are heard only by those who have ears to hear. If there are hours when the immortal hope flickers and burns low in the heart, then must man feed his hope by asking help from earth's wisest spirits, speaking in their noblest moods—for the one-talent man understands himself only in the ten-talent man. Man's rude speech, when developed, becomes eloquence; blundering hands will become skillful; the dull mind may glow and blaze with light; and the children of ignorance and limitations can fully understand their powers only by beholding themselves in the mirror of great men, as in a glass. For every one-talent man is a germinal ten-talent man.

Nature is not the seed when it is planted, but the seed when it is unfolded into the sheaf. Is the wild grape nature, or is nature that grape after it has gone up unto the Concord or the Catawba ? Is the scrub oak nature ? Is not nature the vast acre-covering oak? Is the wild rose nature, rather than the rose made tame, double and of varied hue and sweet perfume ? Nature is not the child Titian—his touches rude, his drawing wrong; but Titian the man, made full by culture—his brush full of ease as the breeze of summer, and full of color as the summer itself. Man suspects not what he is to be until, through growth, his reason works freely; until his speech is perfect, his judgment unerring, his conscience true as the needle to its pole and the heart exhales benefactions as the summer exhales harvests. Just as the young sculptor

86

cannot understand himself until he hath studied Praxiteles; so the soul cannot fully interpret the meaning of the inner voices that whisper their immortal messages until it hath questioned earth's greatest men as to their thoughts and hopes of the life beyond death.

Hours of doubt and denial there may be. Every life knows moments when remorse makes earth a prison; when the sky becomes brass and is let down upon man's forehead; when earth's fruits turn to ashes and soot; when the horizon closes in until man strikes his bleeding knuckles against it as against a wall and the soul is the condemned prisoner of conscience. But repentant hours are followed by hours of aspiration, when the mind becomes luminous and the body walks but does not touch the ground; when the earth seems a cup filled with the wine of life, the heart rises like a bird and God grants such intimations of victory and royalty that the soul walks the earth like a crowned king. In such moments the soul despises arguments of immortality. He who holds a bunch of crimson roses in his hand needs no botany to tell him that roses are beautiful. When the birds are in the hedges and the wheat is in the shock, man needs no almanac to see that it is summer.

So there came to Dante hours when he walked the hills of Paradise, moments when Milton numbered the hosts of the immortal company and Pascal knew the grandeur of the upward flight. With Cicero, confessing that "presentiments of immortality are most discoverable in the greatest geniuses and the most exalted souls," let us question earth's greatest points as to their outlook upon the immortal life.

When an English author asked Phillips Brooks to mention the five mountain-minded men that America had produced, among other names he mentioned the name of Emerson. Fortunately for us, the Sage of Concord lingered long over his study of the immortal life. In the sacred hour of friendship Emerson said: "The resurrection and the continuance of our being is granted. We carry the pledge of this in our own breast. I maintain merely that we cannot say in what form or in what manner our existence will be continued." Later on, when Mr. Emerson felt that his sun was sinking toward the horizon, he reaffirmed his faith. In his last essay, which was a study of immortality, he harvested the full fruitage of his thought and life: " Man is to live hereafter. That the world is for his education, is the only sane solution of the enigma. The planting of a desire indicates that the gratification of that desire is in the constitution of the creature that feels it. The Creator keeps his word with us all. What I have seen teaches me to trust the Creator for all I have not seen. Will you, with vast pains and care, educate your children to produce a masterpiece and then shoot them down ? "

Having affirmed that the soul did not begin when the body began, Mr. Emerson also affirms that the soul is not slain when the body is slain. Foreseeing the end of his career, the sage said: " On the borders of the grave the wise man looks forward with equal elasticity of mind and hope—and why not, after millions of years, on the verge of still newer existence ? I have known admirable persons without feeling that they exhaust the possibili-

ties of virtue and talent. I have seen . . . **what**
glories of climate, of summer mornings and
evenings, of midnight sky! I have enjoyed the
benefits of all this complex machinery of arts
and civilization and its results of comfort!
But the Good Power can easily provide me
millions more." Thus the pages of Emerson
exhale the faith of immortality. Like a cool
spring in the forest, hope bubbles in his
heart. To the Sage of Concord every flying
ideal was God's pledge and promise of a
realm where ideals shall be overtaken and made
man's rich possession. Our author can, indeed,
be quoted against himself, but on the whole,
than Emerson no writer has done more to
strengthen the faith of immortality.

If Emerson grounded immortality upon the
goodness and moral reasonableness of God,
Channing felt that immortality was the log-
ical inference of man's fragmentary development during his early career. Going into the
fields, the scholar noted that once the peach
or pear tree had borne leaves, blossoms and
fruit, its highest end had been fulfilled. Both
root and trunk had, through ripened **fruit,**

touched their climax; and, though centuries many and long were to sweep o'er the fragrant orchard, the years could bring the tree to this alone—leaf, blossom, ripened fruit. But if a few years enable the tree to exhaust its every power and fulfill Nature's every pledge, for man, made in God's image, fourscore years hardly avail to grow the root of industry, much less to exhaust the latent powers of reason or memory or morals. No inventor, like Stevenson, ever had time to work out a tithe of his inventive thoughts. Coleridge left the outlines of severa hundred volumes incomplete. Even Michael Angelo had to divide his life of ninety years into three periods, giving one period to architecture, one to painting and one to sculpture. In his old age, having given a lifetime to poetry, Tenny·son expressed the desire for a like period for music, and similar epochs for science and art and history.

When Coleridge had unfolded his scheme of universal culture to Leigh Hunt, the poet said, sadly: "That means a thousand years in college." When Southey knew that he must die he asked to be carried to his library. There the old scholar went wistfully from book to book, handling each like a dear friend and bidding each a last farewell. For the scholar, separated from his library; for the artist who soon must drop his brush or chisel; for merchant or writer or inventor soon to leave the scenes he loves; for all men with strength, and all women with beauty, there must be written these words: *Too short, the life given!* But God, who gave the tree time to attain its utmost perfection, will not crowd the soul with faculties that demand an eternity for their unfolding, and then cut man off with a brief handful of years. With the gentle Channing, let us believe that for growth the eternal years are ours.

When the stranger knocked at Wordsworth's door and asked if the poet was in his library, the aged servant waved his hand toward the lake and hills and said: "His library is all out of doors." This statement holds equally of our own Bryant. For these are the poets of nature to whom the spirit of the hills and mountains whispered all secrets. The birds,

the trees, the lakes, the white clouds
and perfumed winds; the high hills clad with
forests—these the poets loved, and with them,
as with familiar friends, did linger. Having
observed with what skill the swallow builds its
nest; with what foresight the squirrel lays up
its store against the winter; with what art
the spider spins its web and the wildfowl
finds its way through the pathless air, the poets
have come to believe that if the instinct means
much to animals it means even more to man.
Among the soul's birth-gifts Wordsworth in-
cluded the instincts of God and immortality.
Man comes "trailing clouds of glory." Not
education, not revelation gives the instinct
of an immortal life, but God bequeaths it.

It was this inner voice that whispered to the
ancient Roman that God is immortal and bade
him carve a marble tomb so beautiful that
it seemed not "so much hiding-places of that
which must decay as voluptuous chambers
for immortal spirits." That inner voice also
led the Greeks to ask if they might be buried
where "the sun could see them, and that a
little window might be cut in the tomb from

which the swallow might be seen when it comes back in the spring." Because, for animals, instinct is God's guidebook to the art of living, and has never deceived robin nor butterfly, the poets felt that for man it was safe to trust the instinct of immortality. Therefore, when Bryant saw the water-fowl pursuing its way through the rosy sky he exclaimed:

He who, from zone to zone,
    Guides through the boundless sky thy certain flight,
In the long way that I must tread alone
    Will lead my steps aright.

To the suggestions of poet and philosopher must be added the thought of the scientist. If the time was when Science doubted or denied, now Science has begun to soar with seraphs and to see with saints. Because its instruments are the microscope and the scalpel, physical demonstration is impossible, and Science can neither disprove nor affirm. Yet daily, evolution is unfolding new suggestions and discovering strange analogies and intimations of a life beyond death. The biologists have traced for us the story of the ascent of the human body. For ages hath Nature been toiling upon the perfection of the hand and the foot and the ear and the eye, and these are now well-nigh perfect. At last Science affirms that on earth there will never be a higher creation than man.

The goal toward which Nature hath worked hath been reached, and in developing the mind, Nature is confronted with a stupendous crisis—the arrest of the body. Once man strengthened his eyes by focusing them upon stars distant and great, and also upon crystals near and tiny. Now the field-glass for the distant ship and the microscope for the tiny crystal have arrested the growth of the eye. Once man hurled his spear or held his plow. Now the developments of tools have caused the trip-hammer to succeed the arm and the bicycle to outrun the foot. The mind hath invented a thousand instruments that now fulfill the duties of the body and hath arrested its growth. Herbert Spencer named Romanes as the disciple who had most thoroughly studied the problems of mind from the view-point of evolution, and mentioned John Fiske as the ablest exponent of the general principles of his synthetic philosophy. But Romanes, moving on from higher to higher, came at last to believe that the evolution of the mind involved the final outgrowth of the body and necessitated the casting' it off as a physical clog no longer helpful. John Fiske also affirms that immortality is the one mighty goal toward which nature has been working from the very beginning of life.

"Does death end all?" asks the philosopher. "Has all this work been done for nothing? Is it all ephemeral, all a bubble that bursts, a vision that fades? On such a view the riddle of the universe becomes a riddle without meaning. The more thoroughly we comprehend that process of evolution by which things have come to be what they are, the more we are likely to feel that to deny the everlasting persistence of the spiritual element in man is to rob the whole process of its meaning. For my part, therefore, I believe in the immortality of the soul, not in the sense on which I accept the demonstrable truths of science, but

The Witness of Great Men to Immortality

as a supreme act of faith in the reasonableness
of God's work."

Thus we see that Science also has become a
prophet of faith.

Centuries ago Socrates affirmed that immortality was necessary to reward the good who
have offered their whole lives as a sacrifice for
home and country and progress. But it re-
mained for Martineau to develop the thought
that the protracted life of good men and bad
alike would be fatal to the progress of civilization. Strange that the earthly death of good
men and great is one of God's chiefest boons
to society ! What would be the result if, in the
world of science and letters, great men lived
on for centuries ? Give Newton two hundred
years for astronomy and he will make a complete map of the heavens, search out all laws,
squeeze all the truths from the stars and leave
to young astronomers only a worn and beaten
track. In literature also two hundred years
would enable Scott or Dickens to fill all the libraries with books—tell the story of each generation and century through some noble volume. In the realm of invention, also, if Edi-

son at eighty years of age could begin afresh
and go on for another century, all discoveries
would at last be concentrated in his hands. In
the realm of wealth two generations with two
hundred years each would make all society vas-
sals to a few families. The young need the atmosphere of opportunity and the stimulus of
the unknown. But overshadowed by these
enormous aggregations of wisdom and wealth
and power, young men would shrivel and finally
perish away. Under such conditions the new
ideas of youth could only be introduced by an
earthquake shock or a revolution. And if the
continued existence of the good and great
would be so disastrous, what could be said if
the reins of government were placed in the
hands of some Nero or Napoleon, and continued
there for some two or three hundred years,
during which time men would forget their tra-
ditions of freedom and the ambitious General
would organize the forces of imperialism to
strangle personal liberty ?

Under such conditions free institutions
would become impossible. Therefore, God ordained death "to wrest the incubus from the

breast of dying nations." Better a thousand times a short-lived generation, even though it involves the death of the good, not less than the bad—for the memory of the wicked will soon perish, but the influence and memory of the good and great abide an imperishable stimulus to progress. Because one good custom or man can corrupt a world, and one bad person debase it with centuries of power, death comes in to withdraw the clog and make possible all social progress. Thus the problem is not that man dies so soon, but that man lives so long.

Centuries ago Plato expressed the hope that at some future time the moral law might become a person; that, beholding, all mankind might stand amazed and entranced. Law alone was an abstractum too cold to kindle the heart's enthusiasm. Fulfilling this desire, Jesus Christ entered the earthly scene. He came to teach the disciples of Socrates that nothing evil can befall a good man after death. He came to fulfill the thought of Cicero, that ideals are overtures of immortality. He fulfills Bryant's hope that he who notes the

sparrow's fall will guard his children's graves.
To Tennyson, falling on the altar stairs, that
slope through darkness up to God, he whispers
that for life and death alike there is "one
law, one element, and one far-off divine event
to which the whole creation moves." These
who cast their flowers and tears upon the
grave are bidden to look up and cherish the
memory of the dead, for the friendships begun
in time shall wax through eternity.

Therefore the musician may die to the music of
his own requiem; the poet may pass away to the
note of his own bugle-call; the hero and patriot need not fear when the sunset-gun
doth boom at last. In the gallery of the Vat-
ican the pilgrim reads upon one side the Christian inscriptions, copied from the catacombs,
while on the other side are inscriptions from
the Roman temples. There a single sigh
echoes along the line of white marble: "Farewell, farewell, and forever farewell," But
upon the other side are these words: "He
who dies in Christ dies in peace and hope."
For the hope of immortality is the very genius of Christ's mission and message. God

lives, Christ loves, goodness is eternal; therefore man shall be redeemed out of sin and
death. He who goes down into the grave
is as one who goes down into a great ship
to sail away to some rich and historic clime.
But a divine form stands upon the prow, a
divine hand holds the helm, a divine chart
marks out the voyage, a divine mind knows
where the distant harbor is. In perfect peace
the voyager may sing:

> For though from out our bourne of time and place
>   The flood may bear me far,
> i hope to see my pilot face to face
>   When I have crossed the bar.